Following page. California poppies, showing their propensity for growing in dense, monotypic stands. *Eschscholzia californica*

THE BEST SPRING EVER

WHY EL NIÑO MAKES THE DESERT BLOOM

PHOTOGRAPHY BY
CARLL GOODPASTURE

TEXT BY
JANICE EMILY BOWERS

CALIFORNIA NATIVE PLANT SOCIETY
SACRAMENTO

Camouflage-patterned, night-flying noctuid moths rest during the daytime in the flowers of an albino form of Mojave aster. *Xylorhiza tortifolia.*

All photographs © 2004 by Carll Goodpasture
Text essay © 2004 by Janice Emily Bowers
All Rights Reserved
First Edition

Steve L. Hartman, Editor
Rose Houk, Copy Editor
Melanie Symonds, Illustrations
Thanks to Artworks of Pasadena

Printed and bound in Hong Kong
through Bolton Associates, Inc., San Rafael, CA

Library of Congress Cataloging-in-Publication Data

Goodpasture, Carll, 1943-
    The best spring ever : why El Niño makes the desert bloom /
photography by Carll Goodpasture ; text by Janice Emily Bowers.— 1st ed.
            p. cm.
Includes bibliographical references and index.
    ISBN 0-943460-44-1
1. El Niño Current—Environmental aspects—California, Southern. 2.
Arid regions plants—Climatic factors—California, Southern. 3. Desert
plants—Ecology—California, Southern. 4. California, Southern—Pictorial
works. I. Bowers, Janice Emily. II. Title.

    GC296.8.E4G66 2004
    581.7'54'097949—dc22   2004019325

The California Native Plant Society

is an organization of laypersons and professionals united by an interest in the plants of California. It is open to all. Its principal aims are to preserve the native flora and to add to the knowledge of members and the public at large. It seeks to accomplish the former goal in a number of ways: by undertaking a census of rare, endangered, and extinct plants throughout the state; by acting to save endangered areas through publicity, persuasion, and on occasion, legal action; by providing expert testimony to governmental bodies; and by supporting financially and otherwise the establishment of native plant preserves. Its educational work includes: publication of a quarterly journal, *Fremontia*, and a quarterly *Bulletin*; assistance to teachers and school projects; meetings, field trips, and other activities of local chapters throughout the state. Non-members are welcome to attend meetings and field trips.

The work of the Society is done by volunteers and staff. Money is provided by the dues of members and by funds raised by chapter activities. Additional donations, bequests, and memorial gifts from friends of the Society can assist greatly in carrying forward the work.

Publication of this book was made possible by financial support from the Lois B. and Rudolph E. Fink Publication Fund. Lois and Rudolph were native Californians with an abiding interest in California native plants, particularly in their native habitat. It is hoped that publication of this book will enable others to better experience that same special joy that added so much to the lives of Lois and Rudolph.

Brittlebush goes dormant during the summer dry season, sometimes reduced to a collection of gray and leafless sticks. Winter rains break dormancy, encouraging stems to grow, leaves to sprout, and flower buds to develop. The plentiful flower heads and bushy form of this brittlebush in the Chocolate Mountains, California, bespeak a wet winter. In another month or two, house finches will perch on the plants and eat some of the thousands of seeds. Other seeds will drop to the ground and fall prey to ants. Many seeds, gently buried by wind or animals, will remain to germinate during the next wet winter, which might well be an El Niño year. *Encelia farinosa.*

Preface

Even though I didn't know it at the time, El Niño had a hand in triggering my interest in the desert. It was the spring of 1973 and my brother, a friend, and I drove out to the Tehachapi Mountains where there were flowers everywhere, and then south into the desert where goldfields covered as far as the eye could see east of Lancaster. Not knowing that the fantastic flora was due to unusually heavy rains, I thought that I should plan on visiting the desert "every year" to see the wildflower profusion. Ensuing years revealed that wildflowers in the desert are not an annual occurrence.

Turn the clock forward twenty-five years and I found myself leading a California Native Plant Society (CNPS) desert field trip to Imperial and Riverside counties in what was the "first" advertised El Niño year. It started in the fall with the remnants of a huge Mexican hurricane sweeping northward and dumping inches of rain over all of southern California. Then came a steady series of storms dropping nearly thirty inches in Los Angeles and lesser but significant amounts in the desert. The result was not only many, many annual wildflowers, but larger ones as well, each with many blossoms.

Fortunately, Carll Goodpasture had joined CNPS for this field trip and was capturing one glorious image after another in his inimitable style. We visited desert washes, the Algodones Dunes, and then drove across the Bradshaw Trail.

In nearly every year since, Carll and I have camped in the desert together. During our long drives we began to discuss the concept of a book that would feature Carll's fine images. We then contacted Janice Emily Bowers, author of a number of natural history books focusing on the Southwest. I soon discovered that Janice is a transplanted Californian and a CNPS member. She was inspired by our vision, and thanks to the California Native Plant Society, this book is the result of our collaboration.

While El Niño's most visible characteristic is gorgeous wildflower displays, its impact on wildlife—from insects to reptiles to birds and mammals—is significant. With its biological engine running on high speed, the ecosystem seems to be replenished, the seedbank restored, and desert life booms while the getting is good.

Thanks to El Niño.

—Steven L. Hartman

Ocotillos can gain and lose leaves as many as five or six times a year. After substantial rains, clusters of four to a dozen leaves sprout along the stems. A leaf crop lasts a few weeks or months, until the soil is dry, then drops. Photosynthesis during these brief but frequent periods of leafiness allows plants to manufacture and accumulate enough sugars to support massive flowering after all but the driest winters. *Fouquieria splendens.*

The Way It Is

Day is meaningless without night, the palm of a hand
without the back, the crest of a wave without the trough.
To appreciate a good spring in the desert, the kind
when color dances across the ground and dazzles the
eyes, it helps to know the desert as it is the rest of the
time, during those long months of drought and silence,
months that stretch into years sometimes, into long
periods when a drive across the desert seems best done
quickly if at all and wildflower seeds wait with vegetable
patience for their sixty days of glory.

Say that it is early July in a drought year. No rain
has fallen since the previous January. It's midday, and
the temperature is 110 degrees Fahrenheit. As you step
out of your car, the heat stings your bare face and
hands. Your entire body sags. Around you, the desert
seems to have suspended operations. No lizard skitters
across the sand, no house finch twitters, no ground
squirrel barks in alarm. At your feet, white bursage is
nothing more than a skeleton, having lost all its leaves
some months before. Creosote bushes, with greater
strength of purpose, retain a portion of their leaves, but
these are bronzed instead of green and look as though
they will not last for much longer. Kangaroo rat bur-
rows are blocked by spider webs; clearly no animal has
been in or out for quite some time. Like the dog that
didn't bark in Arthur Conan Doyle's tale, they tell a
story silently.

It takes a prolonged drought—several consecutive years—to
keep ocotillos from blooming in March and April. This reliability
makes them a crucial source of nectar for migrating hummingbirds. That hummingbirds arrive every year in spring makes
them equally reliable as pollinators of ocotillo flowers. Carpenter
bees—sleek black bees about the size of marbles—were once
regarded as robbers of ocotillo nectar because they bite into
the floral tube and extract nectar through the slit. Biologists
have since discovered that carpenter bees move as much or
more ocotillo pollen from plant to plant as hummingbirds.
Where hummingbirds are scarce, ocotillos need carpenter bees
for good fruit and seed production.
Fouquieria splendens.

OF COURSE, deserts are dry. Everything in a desert testifies that aridity is the basic condition. The dark patina on the stones and the paucity of running water; the readiness of leaves to abandon their moorings and the small size and general toughness of those that remain; the absence of large trees and the wide spaces between the shrubs: all this is evidence of a climate that is stingy with moisture, and not only stingy, but unpredictably so. That's the second signal characteristic of deserts—not just the lack of rain but its unpredictability from year to year. The drier the area, the more unpredictable rain becomes.

What can be predicted is that intervals between rains will be long. Over many generations of natural selection, most plants and animals that inhabit the desert have adapted to this. If you live in the desert, you adapt, too. When one sunny day inevitably follows another, and it seems as though the sky has forgotten how to rain, you accept it because the inside of a teacup would not exist without the outside. This desert—the overheated, undershaded expanse of glare and rock that you can hardly wait to escape—this desert is the necessary precondition for the other one, the wildflower garden for which you yearn.

How long must you wait? Not so very long—just about six or seven years, give or take a few.

Lilac sunbonnet, a diminutive annual, does not usually make massive displays but grows scattered among other wildflowers. The distinctively spotted petals distinguish lilac sunbonnet from its close relative, bristly langloisia. The spots guide pollinators to nectar deep inside the floral tube. *Langloisia setosissima* ssp. *punctata.*

Bristly langloisia can be abundant after wet winters, growing in wide-spreading mats comprising many individual plants. When wet, the seed surface becomes mucilaginous. This glue-like substance binds soil grains to the seed surface, making a sheath that retains moisture, thus promoting germination. *Langloisia setosissima* ssp. *setosissima*.

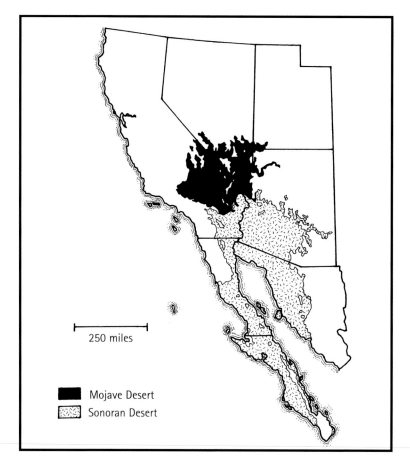

250 miles

■ Mojave Desert
▒ Sonoran Desert

The Mojave Desert

- *Location*: southeastern California, northeastern Arizona, southern Nevada, southwestern Utah
- *Size*: 35,000 square miles
- *Elevation*: 480 feet below sea level to 4900 feet above sea level
- *Physiography*: igneous, metamorphic, or sedimentary surface rock; rugged desert mountains, alluvial fans, dunes, salt pans; interior drainage, permanent springs
- *Climate*: 1 to 8 inches of rain annually, mostly in winter; winters cold, with temperatures dropping to 15 degrees Fahrenheit in valley bottoms; summers very hot
- *Signature plants*: creosote bush, white bursage, Mojave yucca, Joshua tree

The Sonoran Desert

- *Location:* southern Arizona and southeastern California in the United States, Sonora, Baja California, and Baja California Sur in Mexico
- *Size*: 116,000 square miles
- *Elevation:* 180 feet below sea level to about 4900 feet above sea level
- *Physiography*: predominantly igneous and metamorphic surface rock; rugged desert mountains and wide valleys; exterior drainage, little permanent water
- *Climate*: 4 to 12 inches of rain annually split between winter and summer; winters mild, seldom dropping below 25 degrees Fahrenheit; summers very hot
- *Signature plants* : foothill paloverde, saguaro, creosote bush, triangleleaf bursage

Wildflowers on Parade

Nature happens at such a large scale or such a tiny one that most of us never notice it in the course of an ordinary day. Something spectacular—a herd of bison grazing in the distance or a bull moose standing in a river—is needed to grab our attention. Ordinarily, plants can't do it unless they are on fire. Other than forest fires, the only occasion when I have seen large numbers of people look at plants was in spring 1998, one of the best in living memory, a year when the desert seemed to sink under the weight of wildflowers. Many folks stood shin-deep in wildflowers that year, paging through field guides in the hope of identifying the players in that cast of thousands.

About 3,500 species of plants grow in the Mojave and Sonoran deserts altogether. What proportion of these are wildflowers is anybody's guess—in botanical terms, the word "wildflower" is undefined, therefore can mean anything you want. A cactus can be a wildflower, if you like, and so can a shrub. A friend of mine likes to make a distinction between wild flowers (two words) and wildflowers (one word). The former refers to any native plant in bloom, the latter to annuals, which, as every gardener knows, are plants that germinate, flower, set seed, and die in a single year. Winter annuals—in theory if not always in practice—germinate between October and March, flower between February and April, and disperse their seeds shortly before succumbing to hot weather in May.

Some 350 to 400 species of winter annuals can be found in the deserts of California, about half that many in Arizona. Many—perhaps forty percent— would not draw a second glance except from a botanist. Their tiny flowers are white or green. You wouldn't notice them as you sped by at sixty miles an hour, or even at ten. Like plankton in the ocean, they are inconspicuous but vital to the function of the entire ecosystem. Without the Indian wheats, the comb burs, the peppergrasses, and a few other species, the desert would come to a halt.

IT IS TRUE ENOUGH, however, that when most people picture spring in the desert, they paint with vivid colors, pulling from memory a full-page spread in *Sunset Magazine* or *Arizona Highways*. Inconspicuous winter annuals can carpet the desert with green, but they do not make hillsides shimmer with gold and blue, or magenta and purple. That requires bigger blossoms and more color. "Showy" is the botanical term for these species.

Even among showy wildflowers there are gradations in size and color, therefore in overall impact. Of the 200-odd species that qualify as showy, fifty percent are no more than three-eighths of an inch in diameter, about the width of a woman's fingernail or less. Many of these make a splash despite their small individual size by massing together in bunches. About twenty percent are five-eighths of an inch or larger—approximately the size of a nickel. Only a very small proportion, no more than five percent, are greater than one and one-half inches wide. That's about the same diameter as the cap on a jar of fruit juice.

Just as three-eighths to five-eighths of an inch is the most popular diameter, yellow and white are the most popular colors. Thirty-three percent of desert wildflowers are yellow, twenty-nine percent are white. The next largest group is the purples and lavenders at seventeen percent, then the pinks at eleven percent, the blues at seven percent, and finally the oranges at three percent.

These varied colors have meaning that is not apparent to human eyes.

Many diminutive annuals are referred to as "belly flowers" because you have to lie on the ground to see them properly. Here a few have been brought to eye level. Even in a dry year, a careful search will discover wildflowers such as purple mat (top left) and popcorn flower (bottom left) in washes and shady spots.
Nama, Cryptantha, and others.

Like this desert velvet, a large proportion of desert wildflowers are yellow.
Psathyrotes ramosissima.

Yellow flowers as a general rule attract a wide variety of pollinators and are constructed very simply so that pollen and nectar are readily available. Blue flowers seem to appeal particularly to bees, and many are constructed so that bees are the only visitors that can get inside where the goodies are stored. White wildflowers that open around dusk are typically pollinated by moths; the pleasant lemony-sweet fragrance serves as an attractant, and the white color makes them easier to locate in the dark.

Other features that delight us in wildflowers also perform very specific functions. Many wildflowers possess visual cues that direct bees and other flower visitors to small glands inside the blossom. These glands, called nectaries, secrete sugar-rich fluid, a good source of energy. In approaching them, the insect often ends up carrying pollen from the stamens of one flower to the pistils of another, serving as a pollinator in exchange for a nectar reward. One example of nectar guides are the dark pink dots in the center of a pale pink five-spot flower. The contrast between the center of the flower and the outer edge makes a visual path that leads visitors to the nectaries. Such adaptations are not peculiar to desert wildflowers, of course, but are found worldwide in insect-pollinated flowers.

Those of us with a bias for the spectacular might be disappointed by the predominance of yellowness, whiteness, and smallness. But our color preferences matter not a whit as far as the flowers themselves are concerned. Pleasing pollinators, not admirers laden with Nikons and Canons, is what this show is all about.

Many flowers contain nectar, a food that bees and other insects value highly. A tiny insect could squeeze inside the flowers of Great Basin blue sage to get at the nectar but it probably would not touch the anthers or pick up any pollen. This anthophorid bee is too big to fit inside the flower but can extract nectar by inserting its long tongue into the floral tube. Note that the stamens and pistil poke well outside each individual blossom. As the bee clutches the edges of the flower, its abdomen rubs some pollen on the pistil and picks up more from the stamens. In serving as an inadvertent pollinator, it gets a nectar reward. *Salvia dorrii.*

Eyewitness Accounts

An old-timer of my acquaintance, a genial man with a sharp mind and an observant eye, told me that 1941 was the best year ever for wildflowers in southern Arizona. This was in 1979, no mean year itself in my opinion. I remember well how two friends and I drove through endless fields of wildflowers that spring. Because we kept stopping to gawk, we made ninety miles the first day, sixty-five the second, forty-four the third. We buried our faces in wildflowers until yellow pollen smeared our hands, our noses, our eyebrows. I took a photograph of one friend sprawled on his back among sand verbenas and evening primroses. His legs are spread, his arms are flung wide: he has surrendered.

Not a bad year at all.

Who can figure out how that particular year compared to 1941, the year when my old-timer saw endless fields of poppies? And is it even fair to compare them? I like poppies, no doubt about that, but to my mind a special year must be a kaleidoscope of color. I'm greedy enough to want poppies and lupines, or, preferably, poppies, lupines, owl's clover, and a dozen more besides. That is what I yearn for in those long, disheartening dry spells that I try so hard to accept.

The same friend who draws a distinction between wild flowers and wildflowers has rated half a century of springtime displays. His categories—poor, fair, good, great, and spectacular—pretty much cover the spectrum, but, as he himself admits, it is not a perfect system. It relies on eyewitness accounts—mainly his own—and the years before he moved to the region are mostly a blank except for 1941, which he learned from that same old-timer to regard as a very special year indeed. Another problem with his system is that no two pairs of eyes see the same spring, and while I am on the phone to relatives in Iowa, urging them to gas up the motor home and head on out, my friend might be e-mailing professional photographers and telling them not to bother, better luck next year.

Not many kinds of wildflowers thrive on desert pavement, those empty plains where darkly varnished pebbles fit together like puzzle pieces, but desert star seems to prefer that difficult habitat. Like many wildflowers, desert star cuts its coat to suit the cloth, producing many flower heads after wet winters and just one or two in drier years. *Monoptilon bellioides.*

FORTUNATELY, there is a way around these difficulties. Botanists have worked in the Mojave and Sonoran deserts for more than a century now, and for more than a century they have worked in much the same way, fanning out across the landscape in all seasons and years, happily cramming wildflowers into overflowing plant presses, and disgorging the results into herbaria as pressed specimens. Every specimen has a label, and on every label is the place and date of collection. It is a simple matter, then, to make a list of common and showy desert wildflowers, then go through the herbarium to find out which were collected during each year of the twentieth century. Only a few species on the list are likely to be collected in a poor spring, but in a better year, a much higher proportion will turn up, making it possible to guess with some confidence whether a year was better than average and to compare years in a more-or-less objective fashion.

This method is not perfect, either. It lacks subtlety, drawing no distinction between fair and good, or good and great, and it is vulnerable to idiosyncrasies such as gas rationing during the Second World War, when botanists necessarily curtailed their travels. But it is objective, and it lets us peer backward in time beyond the memory of the oldest living inhabitants.

From herbarium collections, then, it appears that unusually good wildflower displays happened about fifteen to twenty times during the twentieth century, or every five to seven years. That's a frequency, of course, not a schedule. In recent decades, 1973, 1978, 1979, 1983, 1992, 1993, 1998, and 2001 saw remarkable shows in some places, if not everywhere. With luck, a person might witness eight or nine better-than-average years over the course of a lifetime. That's much better odds than for Halley's comet, last visible in 1986 and not expected to reappear until 2061, probably even better than for a total solar eclipse, unless you undertake some lengthy and expensive travels.

Desert lily, also called ajo lily, grows from a bulb buried at least a foot underground. After dry winters, the bulbs remain dormant. In years of normal or better rainfall, the bulb first sends up long leaves with characteristically wavy edges, then a sturdy flower stalk with large, fragrant flowers. When winters have been unusually wet, as in many El Niño years, the stalks reach knee height or even taller. *Hesperocallis undulata.*

The Long Sleep

The basic strategy of desert wildflowers is not to adapt to the desert but to escape it insofar as possible. Most of a winter annual's aboveground life coincides with cool, moist conditions. When the weather is hot and dry, winter annuals exist only as seeds in the soil. Seeds thus have tremendous importance as the only link to the future. If all seeds of, say, Arizona lupine germinate at once, and all the seedlings die before they produce any seed, none of us will ever see Arizona lupines again. Seeds need to be hoarded, therefore, and spent cautiously. At the same time, seeds in the soil lose viability or get eaten. The longer they wait, the greater the risk. Caught between an existential rock and a motivational hard place, winter annuals have evolved to play the percentages with the dedication, if not the fervor, of a Vegas gambler. With gamblers, we call it an addiction. With wildflowers, we refer to it as predictive germination.

A seedling, probably desert dandelion, emerges from a crack in the soil surface. Seeds of many wildflowers manage to escape hungry ants and rodents by falling or drifting into cracks like one. *Malacothrix* sp.

Ants find seeds by sight, rodents by smell, but neither group finds every single seed. If seed predators were one hundred percent efficient, desert annuals like this evening primrose would become extinct because there would be no seeds left to carry on the next generation. *Oenothera* sp.

Predictive germination is a calibrated response to environmental conditions. After a very dry winter, seeds of winter annuals do nothing but wait, just like the rest of us, but given a modicum of moisture when temperatures are cool, a small proportion are likely to act. The chance that any of these seedlings will grow to maturity and drop a few seeds is not good, but even a slight chance is better than none—worth a slight risk, at any rate. Thousands of people buy lottery tickets despite equally long odds: to spend a couple of dollars for the chance at several million is probably a waste of money, they tell themselves, but heck, what's two dollars? And so, given a smallish rain, a few seeds will pop their coats, thrust a root into the ground, and poke some leaves into the air. Probably they won't live long enough to reproduce, but the possibility exists, so there they are. Plenty of seeds remain in the soil as a safeguard against extinction. Another year, or later that same season, a somewhat heavier storm will get a bigger response. As the risks go down, the proportion of seeds that germinate goes up.

Seeds of certain winter annuals in our region can lie low for as long as a decade. The means of escape is seed dormancy—a complex state that can involve physiological, chemical, or physical barriers to germination. Some seeds are dormant as soon as they drop from the parent plant; others acquire dormancy as they sit on or in the soil. Either way, a dormant seed by definition will not germinate. Period. Even if you supply the optimum conditions of moisture, temperature, and light, a dormant seed must be released from dormancy before it can respond.

Dormancy is "enforced," as they say, in one of several ways. In some cases, the embryo inside the seed is only part-way grown when the seed is dispersed. Dormancy ends when growth is complete, which requires weeks or months at appropriate temperatures. In other cases, the exterior shell of the seed, its coat, is extremely hard, forming a kind of prison around the embryo. Only after abrasion or high temperature breaks down the seed coat can the embryo escape. Another type of dormancy is enforced by chemicals in the seed that inhibit germination; prolonged leaching as moisture percolates through the soil brings about release. Seeds of desert wildflowers often require high temperatures for release from dormancy, which is not surprising given that the soil surface can reach 160 degrees Fahrenheit or more in summer.

Dormancy comes from the Latin *dormire*, to sleep, which conveys a good idea of the processes involved. A sleeping seed, like a hibernating chipmunk or bear, uses stored food for respiration and for synthesis of the chemicals necessary for life. It is very much alive and very vulnerable. Mice, packrats, ground squirrels, kangaroo rats, and harvester ants eat prodigious quantities of seed. Molds and bacteria devour seeds in an equally implacable if less obvious fashion. Time is an enemy too. As a seed lies in the soil, its water content drops. The longer it waits, the more water it loses, and the more likely it is to experience irreparable damage at the cellular level. At the same time, food reserves are gradually exhausted or chemically altered to useless forms. Eventually the seed loses viability altogether. How long this takes varies according to species and environmental conditions. Seeds of willow and cottonwood die within several weeks of dispersal, whereas seeds of certain weeds can survive in the ground for 100 years. Theoretically, seeds of desert annuals are capable of a very long sleep indeed.

Rigid spiny herb never looks any showier than it does right here. "Showy" is the term botanists use for colorful or large flowers. Like many inconspicuous wildflowers, rigid spiny herb makes up in usefulness what it lacks in display. Plants are often abundant, even after winters that are not especially wet, and the seeds therefore become a reliable and frequently renewed source of food for ants and rodents. *Chorizanthe rigida.*

The Big Awakening

Wildflower seeds in the desert typically become dormant within a few months of dispersal, are released from dormancy by midsummer heat, and are then ready to germinate during the following autumn or winter given suitable temperatures and enough moisture. Exactly how much rain it takes is a remarkably slippery fact. A few species have a minimum requirement of less than half an inch, and a few require a full inch or more. Most are somewhere in the middle. Because small storms are more frequent than large storms, plants with small thresholds tend to appear in more years than those needing big rains.

Keep in mind, however, the elusive nature of these rainfall thresholds.

Wildflowers adapt to local circumstances, and seeds of Indian wheat in the desert near Tucson, where winters are relatively wet, might require twice as much rain for germination as seeds in a much drier location such as Yuma. The temporal distribution of rain matters, too. Half an inch of rain delivered in a single day is satisfactory; the same amount spread across a month is virtually useless. As a general rule, the entire amount needs to fall within one to three days. But not always; sometimes a storm too small to trigger germination is effective if it follows a similar small storm by a week or so.

Two native bees curl up for a good snooze in a poppy flower. They have to sleep somewhere, after all, and often it is in the flowers where they eat and mate. *Eschscholzia* sp.

18

Seeds of winter annuals are just as finicky with regard to temperature. A few seeds might germinate when temperatures are higher or lower than the optimum range, but the best germination occurs within it. A typical range is forty to fifty degrees Fahrenheit at night and fifty to seventy degrees during the day. Because different species have different requirements, the greatest diversity of wildflowers occurs with ample rains over a wide range of temperatures, which means good storms from autumn through winter. Typically, less conspicuous species—the plankton of the desert—are fairly cold-tolerant and can be found germinating in December or January. Many of the showier wildflowers like some warmth and are most likely to appear with autumn rains, should any happen. When temperatures are too cold, they will not respond to even the most generous storms. Wildflowers in the Water-leaf, Poppy, Phlox, Evening Primrose, and Pea families are among this group. Many species in these families have moderate to large rainfall thresholds, three-quarters of an inch or more, which is why the prettiest wildflower displays are associated with the wettest winters.

Even if enough rain falls in late September or October to stimulate seeds of lupines and poppies and owl's clover and whatever else your heart desires, that in itself is not enough to guarantee a spectacular spring. For that, you need frequent heavy rains starting in autumn and continuing all through the winter and maybe into the spring. You need, in fact, at least four inches of rain between September and March in the Mojave Desert, at least eight inches in the Sonoran Desert. In other words, anywhere from thirty-three to fifty percent more than normal for that time of year.

That's a lot of rain for a desert.

A Lot of Rain

And that's where El Niño comes in. Remember 1998 when California rivers from Eureka to Ventura flooded and landslides flattened apartment buildings, destroyed houses, and obliterated highways? That was an El Niño year. The year 1983, when winter storms did at least $10 billion of damage worldwide, one-tenth of that in California alone, was another El Niño year. That time, Arizona rivers flooded, even the dry ones, and the desert city of Tucson became an island for a day or two.

In much of the world, the prediction of an upcoming El Niño year is nothing but bad news. El Niño storms wreck fisheries, level villages, foster epidemics, destabilize governments, kill thousands of people directly and many more through indirect impacts. In addition to this human cost, there is a biological cost, too. Off the coast of California, for instance, mortality of fur seals and sea lions shoots upward because warm ocean temperatures displace their food source, mainly squid and fish, forcing adults to travel so far from the rookery that the pups starve to death before the parents return. Kelp forests off the California coast are vulnerable to El Niño years, as well, because giant kelp cannot survive for long in warm water. Die-off of kelp means loss of animals that depend on kelp forests for shelter or sustenance.

Yet, while El Niño kills ocean-going birds in one part of the world and promotes devastating wildfires in another, it also makes the desert bloom. More than that, El Niño keeps the desert alive.

An El Niño episode begins with interactions between the ocean and the atmosphere in the tropical southern Pacific Ocean. Ordinarily, trade winds push moist air across the tropical Pacific from South America to southeast Asia, gathering more moisture as they go. Upon reaching the warm water and convection currents of the western Pacific, this moisture-laden air ascends. As air rises, it cools; as it cools, it loses its capacity to hold water. The result is copious rain in southeast Asia and, on the other side of the ocean, copious drought.

About every two to seven years, the trade winds weaken. Warm water shifts from the Asian side of the tropical Pacific to the South American side. Along the coasts of Ecuador and Peru, the chilly Humboldt current is replaced by the warm El Niño current. Convection patterns shift, as well. Moist air now rises in the eastern Pacific, bringing rain to Peru and leaving Indonesia and Australia extraordinarily dry. This is El Niño, an aperiodic anomalous warming of the tropical eastern Pacific Ocean.

One measure of these closely linked changes is the Southern Oscillation Index, or SOI, calculated as the difference between air pressure at Darwin, a coastal city in Australia, and the island of Tahiti. Normally, sea-surface pressure is high in Tahiti and low at Darwin, and SOI is generally positive. During El Niño episodes, the pressure gradient is reversed, and SOI becomes negative for a prolonged period, often twelve to eighteen months. Changes in SOI are one signal that climatologists use to predict El Niño events; they also keep an eye on exactly where convection occurs in the tropical Pacific, and they monitor air pressure and temperature on the surface of the ocean and at depth.

El Niño influences climate world-wide by altering atmospheric circulation patterns. In the western United States, a high-pressure ridge located off the west coast determines the track of the polar jet stream, thus of winter storms that originate in the Pacific. Normally, the polar jet stream veers north when it reaches the ridge, steering winter storms across northern California, Oregon, and Washington. Meanwhile, the subtropical jet stream pulls winter moisture across southern California and Arizona, but only weakly. During El Niño years, the high-pressure ridge is displaced farther west, where it splits approaching storms into two tracks. The northern track heads for Alaska, leaving the Pacific Northwest dry. The southern track moves toward southern California and Arizona where it nearly merges with a stronger-than-usual subtropical jet stream. The result can be more fall and winter rain for the deserts, both in terms of frequency and intensity—perhaps two to three times as much rain as in the driest years. The operative word here is *can*, because not all El Niño years are wetter than usual. El Niño merely increases the likelihood of a wet winter to about sixty-five to seventy-five percent.

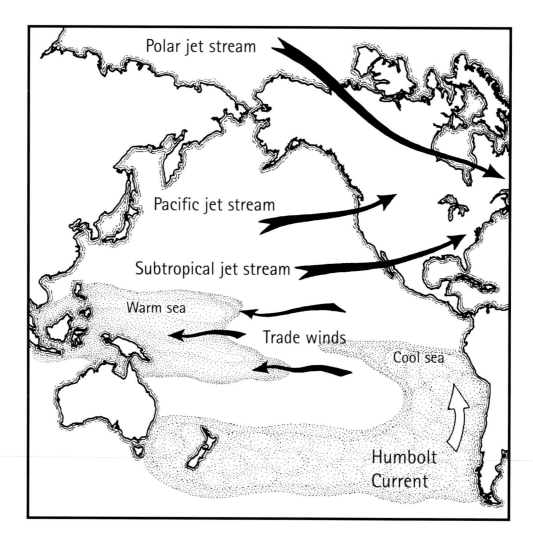

Some major features of ocean and atmosphere in the Pacific Ocean during a normal year. The cold waters of the Humboldt Current maintain cool seas and an arid climate along the Pacific Coast of South America. Trade winds push moist air from the cool seas of the eastern Pacific Ocean to the warm seas of the western Pacific. Moist air rising over the warm water brings copious rain to southeast Asia. In the northern Pacific Ocean, the polar jet stream carries storm systems into Canada and the northern United States but avoids the desert region of the Southwest. Some moisture may penetrate into the Southwest via the subtropical jet stream.

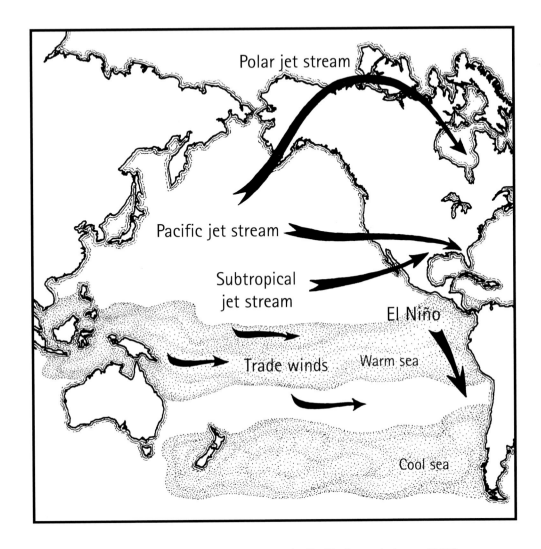

Some major features of ocean and atmosphere in the Pacific Ocean during an El Niño year. Trade winds weaken or even reverse direction, allowing warm water to shift from the western to the eastern tropical Pacific Ocean. As the warm El Niño current replaces the chilly Humboldt current, moist air rises in the eastern Pacific, bringing rain to Peru and leaving southeast Asia dry. In the northern Pacific Ocean, the polar jet stream splits. The southern tongue merges with the subtropical jet stream, carrying Pacific storms across the southwestern United States and increasing the probability of a wetter than normal winter in the desert region.

THE POTENTIAL BENEFITS of an El Niño episode for winter annuals are obvious, so much so that the connection between El Niño and desert wildflowers has become something that everyone "just knows." More satisfying to a scientific mind, the correlation is statistically significant, as well, especially if we define El Niño rather narrowly to mean any year when average SOI from July to December is less than zero. Using this definition, we find that showy displays of wildflowers during the twentieth century were 3.6 times more likely after El Niño years than after other years. By drawing in tropical storms that deliver copious amounts of rain in September and October, El Niño encourages the showiest wildflowers to make an appearance. By bringing in one storm after another throughout the fall and winter, El Niño stimulates germination of a wide variety of wildflowers. By keeping the soil moist, El Niño prevents premature death of seedlings and young plants, giving them a chance to bloom and prolonging the bloom once it starts.

But benefits are not guaranteed. About sixty to seventy percent of El Niño years in the twentieth century were not at all remarkable in terms of wildflower displays. What made these years different was what climatologists call "phasing," the timing of temperature and pressure shifts in the tropical Pacific. Before good wildflower years, the Southern Oscillation Index typically dropped into highly negative values in the late spring and early summer and remained highly negative through the end of the year and into the next, right up until wildflowers started blooming. Before mediocre or poor wildflower years, SOI did not drop as low and did not stay low for as long.

If you want to know when the desert is likely (not certain, mind you) to bloom, keep your eye on the Southern Oscillation Index. A switch to negative values during May, June, or July holds some promise for good wildflowers the following year as long as values remain negative for six to eight consecutive months. Yet even a strong El Niño episode does not necessarily enhance winter rain in the desert, and even good winter rains do not necessarily ensure a great spring.

All predictions are fallible. Wildflower blooms remain essentially unpredictable because of the many variables involved. The likelihood of heavier and more frequent rain is only one of them. Warm, windy weather can cut short a promising show by drying out the soil and triggering premature bloom. A cold snap can retard the growth of seedlings or, under extreme circumstances, kill them. Rabbits, ground squirrels, quail, and various insects can gobble up a promising spring, especially when their numbers are high. Sheep and cows can do even more damage. So much can happen, in fact, that a wise person does not count her buds until they actually bloom.

Not much will be left of this annual once the hawkmoth caterpillar has finished with it, but wet El Niño years supply so much fodder that caterpillars cannot possibly eat every leaf and stalk. Thanks to this abundance, and to the wasps and spiders that prey on caterpillars, enough flowers escape for huge numbers of seeds to enter the seedbank.

A Wider View

In the desert, the repercussions of a wet El Niño year spread widely and resound for a long time. When winter annuals are abundant, for example, the various small animals whose diet consists mostly of seeds and herbs do well, and when small animals do well, the larger animals that feed on them—foxes, coyotes, bobcats, owls, snakes—do well also. An abundance of food is the key. Because years of abundance are more likely than not to be El Niño years, one could say that El Niño keeps the desert alive.

Seed-eating ants and rodents, for example, remove about 300 seeds per square yard of ground per day. The amount of seed in a square yard of desert soil varies widely according to time of year and place—at its lowest, the number might drop to less than 200. You see the problem. The solution is a good wildflower year, when 20,000 to 50,000 seeds might fall on that same square yard of ground. Perhaps ninety-five percent of these seeds are from annual plants, mostly from inconspicuous annuals such as Indian wheat, six-weeks fescue, and comb bur. They don't add much to the splendor of a good spring, but they keep the ecosystem running. Seed-eating animals respond enthusiastically to the largesse of a plentiful seed crop. They depend on hoarded seed to get them through lean times, and when seed is abundant, they collect as much as they can, caching it for later retrieval.

Chuckwallas, large-bodied lizards found in rocky places throughout the desert region, are leaf- and flower-eating vegetarians. When annuals are abundant, chuckwallas prosper. During a good wildflower year, most females might reproduce; in a dry year, none. Desert tortoises feed heavily on annuals, too; they also respond to good wildflower years with higher levels of reproduction. Like chuckwallas, pocket mice and other rodents might not breed in dry years. During severe drought their populations plummet because there are too few young animals to replace adults as they get snapped up by coyotes and owls. Wet El Niño years allow populations to rebound.

An Inca dove nests in a staghorn cholla. The abundant seed crops typical of wet winters
sustain these and other seed-eating birds during dry years when few annuals bloom.

The yellow rays of brittlebush flank yellow or brownish-purple disks. Plants with dark-centered flower heads are common in the low desert near the Colorado River. *Encelia farinosa.*

EL NIÑO YEARS also benefit perennial plants and the animals that rely on them for food. Coyote, javelina, packrat, ground squirrel, and cottontail, for instance, all relish the succulent fruits of prickly pear and the crunchy seeds of foothill paloverde. Both prickly pear and paloverde produce more flowers and set more fruits after wet winters, as do many other desert shrubs and trees. Seeds of some woody desert plants, including brittlebush and triangleleaf bursage, germinate only after heavy winter rains, and in a wet El Niño year their seedlings appear by the thousands. Many are eaten but, thanks to the extra moisture in the soil, a few might manage to survive in protected locations. El Niño at least gets them started. Creosote bush seeds are most likely to germinate in large numbers after heavy rains in autumn, and the likelihood of heavy autumn rains is greater in El Niño years. Here, too, survival is a slim chance; the greater the number of seedlings that get started, the better the odds that a few will thrive.

El Niño is certainly an important source of extra moisture in the desert but not the only one. The Pacific Decadal Oscillation (PDO), a phenomenon much like El Niño, also influences precipitation but over longer times. Negative values of PDO are correlated with droughts that last for ten or twenty years, positive values with periods of

Sunset at Pinto Basin, Joshua Tree National Park, California.

above-average rain. El Niño and PDO can reinforce one another, leading to unusually wet or dry conditions. When SOI is positive and PDO is negative, the result can be prolonged drought. This happened during the 1950s, a very dry decade when good wildflower years were few. Although PDO might have entered another negative phase at the very end of the twentieth century, there is no need to panic. Climatic fluctuations are a normal part of the desert environment, and the plant and animal species that live there are well able to cope with normal variation.

Some species of winter annuals in the desert region have been around ever since the late Pleistocene, as long as 12,000 years ago. Their fossilized remains—mostly seeds and leaves—show up in ancient packrat middens preserved in caves and rock shelters. During the late Pleistocene, annual wildflowers enjoyed wetter, milder winters and cooler summers than today. They grew under pinyons and junipers, not saguaros and foothill paloverdes. Around 8,000 to 9,000 years ago, the climate shifted to dry winters and wet summers. Pinyon and juniper retreated to higher elevations; saguaro and paloverde moved into what is now the desert. Somehow, winter annuals hung on through this unpromising time, and through another climatic shift when El Niño was born about 5,000 to 6,000 years ago, bringing abundant winter moisture for the deserts.

That same flexibility should stand winter annuals in good stead during future climatic shifts. One shift that is already taking place is global warming, perhaps as a consequence of increased carbon dioxide in the atmosphere or perhaps from some other cause. No matter what the reason, the desert is a few degrees warmer now than it was a hundred years ago. Whether the trend will persist is unknown, as is the long-term effect on plants and animals. It is possible that the spatial or temporal distribution of winter annuals in the desert will change as they attempt to track those cool, moist conditions that suit them best; it is also possible that they will stay in place and adapt via altered germination requirements. In the past, Indian wheat has done exactly this, and now plants growing in hotter, drier parts of the desert require less moisture for germination and higher temperatures for release from dormancy than those in wetter, cooler locations. Given their remarkable adaptability to date, it seems likely that most species of winter annuals will continue to bloom in the desert for thousands of springs to come.

Beetle spurge, with narrow leaves clustered toward the tops of the stems.
Euphorbia eriantha.

Threats to Desert Wildflowers

Rain falls on the just and the unjust alike, and when desert wildflowers benefit from a wet winter, non-native annual grasses such as red brome and Mediterranean split grass benefit too. When these alien grasses are abundant, they compete with native wildflowers for space, water, and soil nutrients. The natives often react with stunted growth and premature death. This scenario, if repeated many times, might eventually result in the crowding out of natives by exotics. Fortunately, red brome seeds do not last long in the soil. When a dry winter follows a wet one, red brome seeds die while native wildflower seeds simply wait for a better year. Dry years give natives a chance to regroup and probably keep them from being entirely overwhelmed by the exotic grass. But, because there always seem to be a few locations where red brome manages to reproduce even in a dry year, there is always a source of seed for reinvasion, and the species never dies out entirely.

The biggest threat from red brome is that its dense stands, once they dry up, carry fire much more efficiently than native wildflowers. In some ecosystems—grasslands, chaparral, and pine forests, for example—fire is a natural and even a beneficial process. It thins out crowded saplings, cracks seed coats for germination, and acts as a stimulus for renewal. In these ecosystems, many plants are adapted to fire: in chaparral, shrubs often resprout from the root crown after burning, and, in pine forest, thick bark helps mature trees survive. None of this applies to desert ecosystems, which have not evolved with fire. When fire sweeps through a stand of red brome, most of the long-lived shrubs and cacti are killed outright, leaving barren land. Wildflowers and short-lived shrubs gradually colonize the burned area, which is good, but so does red brome, which only increases the likelihood of more fires in the future. Wet El Niño years, by encouraging thick growth of red brome, speed the transformation.

The desert is a big place, but not as big as it was several decades ago. Between 1950 and 2000, populations of some of the larger towns and cities

Sahara mustard, an Old World annual, appears here as dried, brown stems that have already gone to seed. Invasions often start with just a few plants, then, if conditions are right, accelerate until the exotic is as abundant and widespread as any native. Roads and other kinds of disturbance give exotics a foothold in natural communities. Wet years allow them to build up their numbers. *Brassica tournefortii.*

grew by 500 percent. If the pace of development does not slacken, the next fifty years will bring housing developments right to the edge of desert parks and monuments. Vast tracts of desert that have always been wild will be wild no longer because people necessarily exert pressure on their environment. We need electricity, for example, so we turn "useless" desert into windmill farms. We want recreation, so we drive our vehicles wherever they can go. We generate trash, we requisition water from far away or deep underground, we turn mountains into mines, we bulldoze wildflower fields.

Exotic weeds, like native wildflowers, flourished during the El Niño year of 1998. Here, an annual grass, perhaps Mediterranean split grass, from the Old World, nearly covers the ground. *Schismus* sp.

SOME ANIMALS and plants thrive in the zone where people and desert meet—ravens forage in dumpsters and parking lots, and red brome follows roads wherever they lead. Other organisms, less adaptable, fare poorly. The desert tortoise, the Coachella Valley fringe-toed lizard, and the Mohave ground squirrel are among them. Also on that list are a number of desert annuals, including Coachella Valley milk-vetch, known from twenty sites, Little San Bernardino Mountains gilia, known from ten, and Kelso Creek monkey-flower, known from seven. Altogether about three dozen annual wildflowers indigenous to the Mojave Desert are considered rare, threatened, or endangered in California, if not elsewhere. According to the California Native Plant Society, threats to the continued survival of these and other desert wildflowers include, as you might expect, vehicles, energy development, mining, and exotic plants. The biggest threat, however, is suburban sprawl. Wildflower seeds can survive years of drought, but even the wettest El Niño can do nothing for them once their seeds have been buried under asphalt and concrete.

Twining milkweed,
*Sarcostemma
cynanchoides.*

Sometimes wildflower lovers become wildflower tramplers, as happened here at Anza–Borrego Desert State Park. During the El Niño year of 1998, some 10,000 visitors passed through the park daily, eager not to miss a once-in-a-lifetime event. *Abronia villosa.*

Hotspots

Every spring, the rumors fly: it's going to be a great year for wildflowers, it's not as good as last year, it's pretty good in one place but not at another, it's as bad as it has ever been. . . . What's a person to do? And how can there be so much disagreement about such a simple matter?

In theory, when wildflowers are abundant in Joshua Tree National Park, they should be abundant at Organ Pipe Cactus National Monument, too, because the tropical cyclones and winter frontal storms that trigger germination typically cover a wide area. In the best El Niño years, the entire desert becomes a wildflower hotspot. But this is not always the case. Even a broad storm can vary enough in spatial coverage and intensity that adjacent counties can differ considerably in what kind of and how many wildflowers appear. That makes it difficult to make a list of hotspots, as one might for birders; there is no guarantee that any particular place will always be good.

Fortunately, state and federal parks are eager to accommodate wildflower seekers. Some establish wildflower hotlines where you can call for specific information as to place and time of peak bloom. Many post wildflower updates on their web pages. In Arizona, the Desert Botanical Garden (Phoenix) and the Arizona-Sonora Desert Museum (Tucson) are additional sources of information, as are the Living Desert Museum (Palm Springs) and Anza Borrego Desert State Park (Borrego Springs) in California.

Keep in mind that except in dry years when almost no seeds germinate, you can, with careful searching, find many kinds of wildflowers even if there are no showy displays. The secret is to look in sandy washes and shady spots. But you have to be out there looking, and this is something few people do in an unpromising season.

IN SPRING OF 1998, that never-to-be-forgotten year, my husband and I camped in a desert canyon in Death Valley. It turned out to be a magical time and place. We noticed bighorn tracks in mud and found a bobcat lair marked with fresh scat. We flushed an owl from a cottonwood tree and saw a hummingbird alight on her nest. We found wildflowers we had never seen before and might never see again. At the end of our sojourn, it was harder than usual to depart, but of course with jobs and a cat waiting back home, we had little choice in the matter.

On the homeward drive, we stopped to take a few last pictures where an embankment was piled high with phacelia, desert sunflower, desert lupine, and sundrops. As I buried my nose in the sundrops and breathed in their spicy fragrance, wondering if I could bear to leave, I realized that breaking camp that morning would have been much easier if I had known about this gift just fifty miles down the road. By analogy, then, I knew that I could indeed leave this incomparable spring—this miraculous, unpredictable conjunction of time and place—confident that there would be other springs, equally good, or, if not that, confident that even if I did stay, spring would move forward and away from me, so we might as well go.

And so we did.

A single flower head of
Algodones Dunes sunflower is a
supermarket, supplying many
needs. *Left*, a crab spider lies
in wait underneath the flower
head for unwary insect visitors;
below, a solitary bee scrambles
pollen into her baskets.
Helianthus niveus ssp. *tephrodes*.

Opposite. Algodones Dunes sunflower, like
many perennial plants found on active dunes,
risks constant burial or excavation as wind
moves sand from here to there. Its seeds germinate abundantly after autumn rains, generally tropical cyclones that penetrate farther
north than usual. The likelihood of such
storms goes up in El Niño years. This species
can be found only on dunes and sand sheets
in the western Sonoran Desert.
Helianthus niveus ssp. *tephrodes*.

The large and fragrant flowers of this evening primrose open late in the day and close the next morning. Night-flying moths, attracted by the sweet odor, are the major pollinators. Solitary bees, active during daylight hours, gather up strands of pollen but probably do little pollination. *Oenothera deltoides*.

White flowers show up well by moonlight. On darker nights, moths locate night-blooming flowers by scent. Such blossoms typically smell musky, sweet, or lemony.

At the Living Desert Museum, Palm Desert, California, a gray hairstreak hovers among the long, red stamens of Baja fairy duster, an attractive shrub native to Mexico. The picture, taken in a dry spring, shows how butterflies benefit from our human urge to cultivate flowering plants. *Calliandra californica.*

Purple hairstreak (top left), queen (top right), painted lady (bottom left), and Mormon checkerspot (bottom right) butterflies taking nectar from a groundsel blooming in cultivation at the Arizona–Sonora Desert Museum, Tucson (*Senecio* sp.). Butterflies seem especially abundant during wet El Niño years when wildflowers bloom for many months in succession, providing food for one brood after another. Such years often see vast numbers of painted ladies undertaking one-way, low-level migrations across the desert. Sand verbena is a favorite nectar plant of painted lady butterflies in the desert region. The caterpillars feed on fiddleneck, popcorn flower, mallow, and many other kinds of plants.

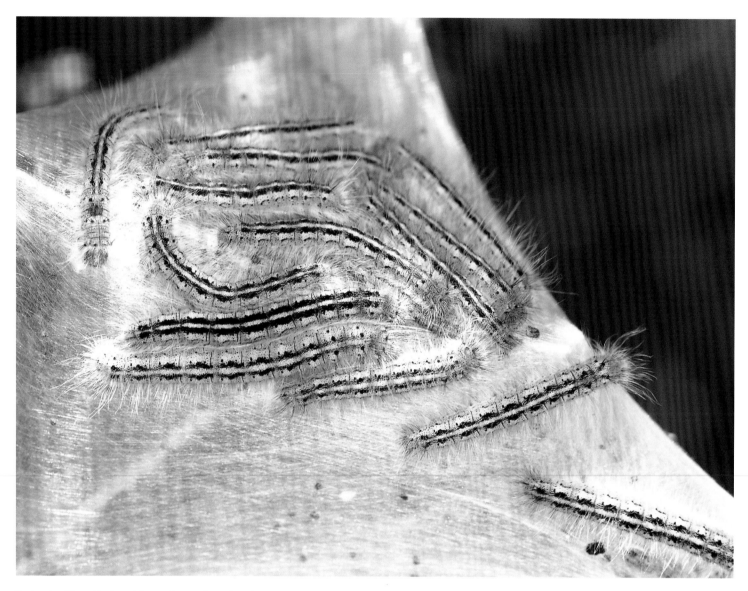

Tent caterpillars, the larval form of a small moth, spin communal webs as daytime retreats, then emerge in unison at dusk to feed on tender spring foliage of shrubs such as saltbush and shadscale.

Opposite. As an adult, the white-lined sphinx moth seeks nectar in fragrant flowers that are open at night. Sphinx moths are important pollinators of evening primroses and various members of the Phlox family. As a caterpillar (inset), this moth feeds on a variety of wildflowers, especially the evening primroses. When abundant, as is often the case after wet winters, sphinx moth larvae are capable of defoliating entire plants.

Jojoba bears male flowers (above) and female flowers (right) on separate plants. Showy petals, which would attract insect pollinators, are not needed because wind carries pollen from male to female flowers. When fertilized, the ovary of the female flower becomes a hard, brown nut rich in wax, a high-calorie source of energy. The nut, which is of course a seed, metabolizes some of this wax during germination, and the seedling uses much of the rest during its first month of life.   *Simmondsia chinensis.*

Jojoba is a long-lived shrub well suited to an arid climate. Leaves are leathery and coated with wax, thus resistant to water loss, and growth is slow, making conservative use of soil moisture. Yet even drought-tolerant plants suffer during prolonged drought. This jojoba, shown in March 2002 after a very dry winter, can no longer support all the leaves it put on during an optimistic growth spurt the previous year. The brown leaves will soon drop. Although March is the season for jojoba bloom, this drought-stressed plant is not flowering.   *Simmondsia chinensis.*

A jojoba under irrigation in March 2002. An adequate water supply makes all the difference, and balls of male flowers dangle from the stems of this jojoba growing at the Rancho Santa Ana Botanic Garden in Claremont, California.   *Simmondsia chinensis.*

Above and opposite, cacti start blooming as the wildflower season winds down. In the Sonoran Desert, various species of hedgehog cacti start the show in April with large, red-violet or magenta blossoms. The stems and immature fruits of Engelmann hedgehog are well protected by spines. When ripe, fruits lose their spines, becoming accessible to ground squirrels, rabbits, and other animals. Consumers become dispersers when viable seeds pass through digestive tracts to be deposited in scats.   *Echinocereus engelmannii.*

Mormon tea, a common desert shrub, shown here on the Algodones Dunes, California, in March 1998. Like pine and fir, Mormon tea is a gymnosperm and produces cones instead of flowers and fruits. Scales of pine cones are hard and fibrous, whereas those of Mormon tea are papery. Looking very much like a cone, a crab spider (above right) suspends itself between two stems of Mormon tea. Camouflage makes the spider nearly invisible to prey and predators both. The trick works as long as the spider remains motionless. *Ephedra trifurca.*

Sand food, one of the strangest of desert plants, is a parasite restricted to dunes and sand sheets in the western Sonoran Desert. Deeply buried stems attach to the roots of host plants, extracting water, sugars, and nutrients. In spring, the underground stems send up flower stalks. The inflorescences, fleshy heads as large as bread-and-butter plates, lie flat on the sand surface. Blooming of the small, pink or purple flowers starts in the center of the head, then radiates toward the edges. Sand food reaches its largest size and greatest abundance when winters have been wet.   *Pholisma sonorae.*

Of several hundred globe mallows growing in a dry wash in Imperial County, California, only this one had white flowers. The rest were orange, as is typical for the species. Simple mutations produce variant flower colors such as white instead of blue, or yellow instead of red. The variations are most likely to be seen when populations are large. *Sphaeralcea angustifolia.*

The wetter the winter, the taller the stems of the parachute plant. Wetter winters also bring more flower heads and larger leaves. The energy required for bolting and blooming can deplete the leaves of carbohydrates, and they often wither and die before the end of the flowering season. *Atrichoseris platyphylla.*

One of many species of phacelia, probably notch-leaved phacelia. Some, like this one, are also called scorpionweed because the curved flower stalk is reminiscent of a scorpion's tail. The botanical term for this type of inflorescence is "scorpioid." Scorpioid inflorescences are characteristic of the genus. *Phacelia crenulata.*

A mat-forming annual, silky dalea has long, soft hairs on the calyx lobes.   *Dalea mollis.*

An anthophorid bee clasps a wolfberry flower as she uses her long tongue to probe for nectar. Except for the bumble bee, desert bees are solitary, meaning that they build and provision their nests as individuals rather than as cooperative members of a hive or colony. Anthophorids are important pollinators in the desert. *Lycium* sp.

Whether in flower or in fruit, the ten species of desert shrubs collectively known as wolfberry or boxthorn fill several different needs. Birds and ground squirrels relish the succulent fruits, bees and butterflies get nectar from the flowers, and birds nest in the dense canopies. Some species of wolfberry have squat lavender flowers, others slender white ones. Substantial storms are required for the best flower and fruit crops. *Lycium* sp.

Desert calico lasts well into the spring, often blooming after other annual wildflowers have stopped. *Loeseliastrum matthewsii.*

A common shrub in the Mojave Desert, Great Basin blue sage has gray foliage and purple flowers borne in clusters on the flower stalks. When crushed, the leaves release the strong aromatic odor characteristic of sages. *Salvia dorrii.*

Orcutt's woody aster, a rare, cliff-hugging shrub found in desert canyons. The flower heads are large for a desert plant, as much as three or four inches across. *Xylorhiza orcuttii.*

Mojave aster, a perennial wildflower, does not come up from seed every year but from underground rootstocks capable of outlasting seasonal drought. Each purple "petal" is actually an individual flower. The yellow center is also made of many individual flowers. *Xylorhiza tortifolia.*

A solitary bee on Mojave aster. The pollen baskets on her legs are nearly full. Male and female bees visit flowers for nectar, but only the female harvests pollen. Her larvae feed on it after they hatch from eggs which she has laid in specially prepared nests. *Xylorhiza tortifolia.*

Blazing star. Its sandpapery leaves are characteristic of the Loasa family to which it belongs.   *Mentzelia involucrata.*

Insects pollinate by carrying pollen from a flower on one plant to a flower on another plant of the same kind. Pollen comes from stamens, the male reproductive organs; to be useful, pollen must be brushed onto the pistil, which is the female reproductive organ. Here a bee has landed on pistils at the center of the flower. Surrounding the bee are numerous stamens thickly dusted with white grains of pollen. With any luck, it has brought blazing-star pollen from another flower and will take away pollen from this flower when it leaves. *Mentzelia involucrata.*

Tiny mesquite flowers, crowded by the hundreds in pendant inflorescences, attract hordes of visitors, especially wasps and bees but also beetles, bugs, butterflies, and the various insects and birds that prey on them. A mesquite tree in full bloom is a city abuzz with varied and hectic street life.
Prosopis glandulosa.

Like mesquite, catclaw has tiny flowers densely clustered into cylindrical inflorescences. Both are woody plants in the Legume family. Catclaw flowers smell like honey, and the nectar can be a major component in the honey made by honey bees. Here large wasps known as tarantula hawks find nectar in the blossoms. *Acacia greggii.*

Female wasps are active hunters and require high-energy food sources to sustain their predatory lifestyle. This sphecid wasp takes a break from hunting to drink nectar at flowers of Mojave stinkweed. *Cleomella obtusifolia.*

Tiny bees mate inside the blossom of California poppy growing at Picacho Peak, Arizona. Note the bulging pollen baskets on the legs of the female. Solitary bees are the most important pollinators of desert wildflowers. These are native bees, unlike the honey bee, brought to the Americas by humans. Solitary bees often meet and mate on flowers. Female bees go to flowers for pollen and nectar, making blossoms a likely place for male bees to search for partners. *Eschscholzia californica.*

Desert gold poppy is one of several hundred annual wildflowers to be found during wet El Niño years. The seeds stay dormant in the soil through winter drought and summer heat, for several years if necessary; only the right combination of cool temperatures and moisture can make them germinate. *Eschscholzia glyptosperma.*

Bladderpod, a kind of wild mustard, is among the first wildflowers of spring in the Sonoran Desert, often appearing as early as February. *Lesquerella tenella.*

In the Mojave Desert, where most rain comes in winter, trixis usually blooms only in spring; but in the Sonoran Desert, where half the rain falls in warm months, the shrubs flower in spring, summer, and fall. *Trixis californica.*

Chuparosa, a bushy shrub of washes and shaded slopes, blooms in February and March during the northward migration of hummingbirds across the desert. Without an abundant and dependable nectar source, humming-birds would be hard put to sustain the long trip from tropical to temperate zones. Chuparosa is one of several perennial plants that meet this need. Chuparosa flowers are well adapted for pollination by hummingbirds. The red color signals the presence of nectar, the down-turned lower lip lets humming-birds hover unimpeded, and the projecting stamens deposit pollen on the bird's head. *Justicia californica.*

Wishbone bush on a cloudy day. The flowers shun the hottest, brightest part of the day, opening in the late afternoon and closing the next morning. In cloudy weather, they might stay open all day long. *Mirabilis bigelovii.*

Even in rather dry years, desert chicory can make a splash along roadsides and wherever a little extra moisture accumulates in the soil. *Rafinesquia neomexicana.*

Not all spiders spin webs to catch insect prey. On the left a crab spider sits and waits for unwary flower visitors. Below, a different crab spider successfully ambushes a moth.
Palafoxia sp. (left), *Chaenactis* sp. (below)

A honey bee takes nectar at the small, purple flowers of desert lavender. As a woody perennial present in the landscape year in and year out, desert lavender provides nectar for honey bees when annuals fail, making it an important food source in dry years. Honey bees, introduced from Europe, have become completely naturalized throughout North America. Unfortunately, they compete with native bees for pollen and nectar, a matter of concern to those who appreciate the diversity of native bees. *Hyptis emoryi.*

Flowers of buckhorn cholla can be dark red, yellow, or some shade in between. The color is constant for an individual plant but varies across the population. *Opuntia acanthocarpa.*

Although its stems are succulent, even a cactus suffers during severe drought, and in an attempt to conserve its resources for survival, will wait until another year to reproduce. Here beavertail blooms exuberantly in a wet El Niño year. *Opuntia basilaris.*

Hummingbirds associate the color red with nectar, and many red flowers are indeed pollinated by hummingbirds. The red blossoms of claret cup, a hedgehog cactus, attract hummingbirds but do not exclude other kinds of pollinators. Most cactus flowers close at night. Those of claret cup remain open, perhaps to allow sphinx and hawk moths a chance to feed and pollinate. *Echinocereus triglochidiatus.*

Sand verbena smells so sweet that even a nose located five feet above ground level picks up the fragrance. The odor draws in lepidopteran pollinators—moths at night and butterflies by day. Of the dozen or two dozen flowers in each head, only a few are likely to produce seed, but all are necessary to make a pollinator-attracting display. The overwhelming abundance of sand verbena in good years suggests that a few seeds per head are more than enough to perpetuate the species with plenty of seed left over for ants and rodents. *Abronia villosa.*

Desert bellflowers, Joshua Tree National Park, April 12, 1998, at the peak of the El Niño floral display. *Phacelia campanularia.*

Further Reading

I am deeply grateful to the authors of the many technical and scientific publications that I used in writing this essay, and I regret that space limitations make it impossible for me to acknowledge my debt to each one. Readers interested in learning more about the natural history of the Mojave and Sonoran deserts might enjoy the following works, a small sampling from many possibilities. Some may be out of print but are well worth seeking out.

Animals

Cornett, James W. *Wildlife of the North American Deserts*. Palm Springs: Nature Trails Press, 1987.

Jaeger, Edmund C. *Desert Wildlife*. Stanford: Stanford University Press, 1961.

General

Alcock, John. *Sonoran Desert Spring*. Chicago: University of Chicago Press, 1985.

Alcock, John. *Sonoran Desert Summer*. Tucson: University of Arizona Press, 1990.

Bowers, Janice Emily. *Dune Country: A Naturalist's Look at the Plant Life of Southwestern Sand Dunes*. Tucson: University of Arizona Press, 1998.

Nabhan, Gary Paul. *Gathering the Desert*. Tucson: University of Arizona Press, 1985.

Philips, Steven J. and Patricia Wentworth Comus, editors. *A Natural History of the Sonoran Desert*. Berkeley: University of California Press, 1999.

Tweit, Susan J. *Seasons in the Desert: A Naturalist's Notebook.* San Francisco: Chronicle Books, 1998.

Zwinger, Ann Haymond. *The Mysterious Lands: A Naturalist Explores the Four Great Deserts of the Southwest.* New York: E. P. Dutton, 1989.

WILDFLOWER GUIDES

Arizona Highways. *Desert Wildflowers.* Phoenix: Arizona Department of Transportation, 1999.

Bowers, Janice Emily. *100 Desert Wildflowers of the Southwest.* Tucson: Southwest Parks and Monuments Association, 1989.

Ferris, Roxana S. *Death Valley Wildflowers.* Death Valley: Death Valley Natural History Association, 1974.

Mackay, Pam. *Mojave Desert Wildflowers.* Guilford: Globe Pequot, 2003.

Munz, Philip A. *California Desert Wildflowers.* Berkeley: University of California Press, 1962.

Quinn, Meg. *Wildflowers of the Desert Southwest.* Tucson: Rio Nuevo, 2000.

Spellenberg, Richard. *Sonoran Desert Wildflowers.* Guilford: Globe Pequot, 2003.

GUIDE TO COMMON AND SCIENTIFIC NAMES
OF PLANTS MENTIONED IN TEXT

brittlebush   *Encelia farinosa*
Coachella Valley milk vetch   *Astragalus lentiginosus* var. *coachellae*
comb bur   *Pectocarya heterocarpa, P. platycarpa, P. recurvata,*
     P. setosa
cottonwood   *Populus fremontii*
creosote bush   *Larrea tridentata*
desert lupine   *Lupinus arizonicus*
desert sunflower   *Geraea canescens*
evening primrose   *Oenothera deltoides, O. primiveris*
Evening Primrose family   *Onagraceae*
five-spot   *Eremalche rotundifolia*
foothill paloverde   *Cercidium microphyllum*
giant kelp   *Macrocystis pyrifera*
Great Basin blue sage   *Salvia dorrii*
Indian wheat   *Plantago ovata, P. patagonica*
Joshua tree   *Yucca brevifolia*
juniper   *Juniperus californica, J. osteosperma*
Kelso Creek monkeyflower   *Mimulus shevockii*
Little San Bernardino Mountains gilia   *Gilia maculata*
lupine   *Lupinus arizonicus, L. brevicaulis, L. shockleyi,*
     L. sparsiflorus
Mediterranean split grass   *Schismus arabicus, S. barbatus*
Mojave yucca   *Yucca schidigera*

owl's clover   *Castilleja exserta*
Pea or Legume family   *Fabaceae*
peppergrass   *Lepidium lasiocarpum, L. thurberi,*
     L. virginicum
phacelia   *Phacelia bicolor, P. campanularia, P. crenulata,*
     P. distans, P. fremontii, P. tanacetifolia, P. vallis-mortae
     and others
Phlox family   *Polemoniaceae*
pinyon   *Pinus monophylla*
poppy   *Eschscholzia californica, E. glyptosperma,*
     E. minutiflora, E. parishii
Poppy family   *Papaveraceae*
prickly pear   *Opuntia engelmannii*
red brome   *Bromus rubens*
saguaro   *Carnegiea gigantea*
sand verbena   *Abronia villosa*
six-weeks fescue   *Vulpia octoflora*
sundrops   *Camissonia brevipes, C. cardiophylla*
triangleleaf bursage   *Ambrosia deltoidea*
Water-leaf family   *Hydrophyllaceae*
white bursage   *Ambrosia dumosa*
willow   *Salix gooddingii*

Harvester spider consuming
ocotillo flower.
Fouquieria splendens.

84

Many desert insects rely on flowers as meeting places. These blister beetles, having gathered on the petals of prickly poppy, nibble stamens and (off-camera) mate. *Argemone corymbosa.*

85

PLANT PHOTO INDEX

Front cover: Composite of Anza-Borrego Desert State Park featuring evening primrose and sand verbena.

Back cover (clockwise from top left): desert mallow, *Sphaeralcea angustifolia*; desert five-spot, *Eremalche rotundifolia*; desert dandelion, *Malocothrix* sp.; and blazing star, *Mentzelia* sp.